Heritage Farming
in the Southwest

■ Gary Paul Nabhan ■

Western National Parks Association
Tucson, Arizona

ABOUT THE AUTHOR
Gary Paul Nabhan

Gary Paul Nabhan is founder of the Renewing America's Food Traditions collaborative and co-founder of Native Seeds/SEARCH. His twenty books have brought him honors such as a Western National Parks Association Emil W. Haury Award, a MacArthur Fellowship, and the John Burroughs Medal for nature writing.

A University of Arizona research scientist, he has assisted numerous cultures in their efforts to safeguard their heirloom seeds and heritage breeds.

Above: Heritage plants are grown and harvested at the Albuquerque Biological Park, Rio Grande Botanic Garden. Heritage animals, such as Churro sheep, are raised for wool at the park's heritage farm in the New Mexico city.

Introduction

I t's a stormy summer day, and large white clouds are billowing up over a rocky ridge lined with terraces where prickly pears, agaves, mission figs, and Sonoran pomegranates grow in neat rows. Below this ridge-side orchard, a small wash flows into a floodwater-irrigated field of mixed crops. It is situated not far from a turkey pen and a perennial pasture where sheep graze. In the field, the fourth generation of farmers to work this particular land kneel down in the moistened sandy soil and show their crops to a few student interns and visitors. They check the maturation of some heirloom crops that sprouted from seeds they planted nearly three months earlier. In the background, you can hear the baas of some Navajo-Churro sheep, as well as the gobbles of black Spanish turkeys that are being herded from the corral to the field, where they will hunt grasshoppers among the maturing crop plants.

Above: The heritage farm at Albuquerque Biological Park, Rio Grande Botanic Garden includes a garden with traditional Southwest crops, a barn, and a corral for Churro sheep.

The heirloom vegetables—chapalote popcorn, amaranths, tepary beans, and an acorn squash—were familiar to the first farmers in the Southwest some 4,100 years ago, but they—like the heritage flocks of sheep and turkeys—are now undergoing a revival in the region. Although the father of the family sees his work as part of a broader revival of heritage farming, he grins as he explains to the visitors that he has not stepped back into "the dark ages," but forward into the future. Unlike the farmers who grew these same crops centuries before, he uses soil moisture sensors to trigger a computerized drip irrigation system that uses solar power to pump water out of a floodwater reservoir. The farmer reminds us why he sows the particular seeds he has selected:

> *"The so-called 'ancient' crops we choose to grow exhibit a lot of variation within this field, so that they respond well to a changing climate. These crops have been continually adapting and re-adapting to our southwestern soils, short rainy season, and extreme temperatures for centuries, if not millennia. In fact, it is when a new hybrid crop has been planted from a single clone or selection that it becomes too genetically narrow to rapidly adapt to global warming, new pests, or diseases."*

On this particular day, there is excitement in the air, for it seems as though the chapalote corn crop is ready to harvest; the smoky brown kernels are past the milk stage, and have dried down enough for their cobs to be picked. Along with the ears of chapalote to shuck, there are bean pods to thresh, voluptuous squash to harvest, a few golden yellow squash blossoms to pluck, and gorgeous red spikes of amaranth seeds to bag and dry. Mouths begin to water and the harvesters carry on some banter among themselves; dreams of a sustainable agriculture, rooted in tradition, begin to take shape, bright and billowy, like the clouds forming above their heads.

There is something both reverently ancient and refreshingly novel about the way these folks have chosen to farm in the dry, but ever unpredictable landscape of the Southwest. Like a growing number of farmers in Arizona, Nevada, New Mexico, Utah, Texas, and Colorado, they have chosen to blend ancient, hardy

Above: Tyrone Laurence, Bernard Enos, and Ervin Crawford, left to right, pile tepary bean plants into a row for a combine to process at the Papago farm on the Tohono O'odham Nation south of Tucson, Ariz. Inset: A Rio Grande Wild tom turkey is one of a flock raised by Timothy Willms, owner of the Talus Wind Ranch in Galisteo, N.Mex. The breed is thought by some taxonomists to be a feral offspring of domestic turkeys once raised by ancestral Puebloans.

seeds with new technologies in order to meet the challenges of today without throwing everything away that has worked in the past. Because the student interns, elder hostel participants, and eco-tourists who help with the week-long harvest better appreciate where our food comes from, they may feel inclined to rally support for the conservation of seeds, breeds, water, and land, along with sustainable, place-based food production essential to our food security. Perhaps this experience will produce future farmers as well—ones who care about a sustainable future and the great agricultural legacy of the Southwest. It is a legacy that historian Edgar Anderson called "one of the most remarkable agricultural civilizations in the entire world…for it produces more food on less water than any other."

These farmers do more than just regenerate their own hardy seeds from year to year; they are also renewing the Southwest's time-tried agricultural traditions.

Above: Corn harvested on the Hopi reservation in northern Arizona is ground for human consumption and also dried for livestock feed. Inset: Tepary beans have been a staple crop in the Southwest for centuries. Below: Hopi blue corn ready for harvest at Hubbell Trading Post in northeastern Arizona.

This is heritage farming, for it involves the sustainable production of food and fiber from native and historically introduced breeds of livestock, and heirloom seeds of vegetables, fruits, and grains adapted to the region in which they are grown. As long-time rural activist Mark Ritchie of Minnesota has said about a particular set of sustainable farms and ranches in the Southwest, "These landscapes not only produce food and fiber, they also provide renewable energy, ecological services, conservation of cultural heritage, social cohesion, and biological diversity."

In the Southwest, we call particular genetic stocks of tepary beans, grain amaranths, squash, or corn "heirloom" seeds, because they have been passed from generation to generation of farmers. Those seed savers have not only selected these heirlooms for their culinary qualities suited to the region's beloved foodways, but have also selected them for their ability to grow well under the arid conditions of the basin and range habitats of the Southwest. Some, like Hopi blue corn, can be planted a foot deep into the moist sand hidden within a sand dune, and emerge to produce eighteen-inch-long cobs some three months later.

Above: Amaranth is a crop grown historically, high in protein and calcium, and having a fiber content three times that of wheat. Aztecs made an amaranth mixture into idols that were eaten in connection with their sacrifices and religious rituals. Inset: A Hopi dry-land farmer weeds a grouping of corn plants in northern Arizona.

A woman stacks corn at a roadside stand on the Hopi reservation.

Others, like the O'odham brown tepary bean, have been known to reach maturity and produce seeds on the soil moisture accumulated during just two downpours—one gully-washer the day before planting, and the next, twenty days into the heat of August!

However, we seldom use the term "heirloom" to describe animals in the same way we use it for hand-me-down seeds, furniture, or jewelry. Instead, we call the equivalent forms of such livestock—from Criollo-Corriente cattle to Navajo-

Top: A member of the Herder family in Hard Rock, Ariz., tends her sheep. The Herders are one of many Diné families that produce Navajo-Churro lamb and mutton for their own consumption and that of their neighbors to support their people's food self-sufficiency. Lower left: A farmer works in a tepary bean field on the Tohono O'odham Nation. Lower right: A Hopi woman fashions piki bread from ground blue corn mixed with water and the burned ashes of native bushes. Spread over a heated baking stone, the bread is coated with oil derived from the seeds of squash, watermelon, or sunflowers.

Churro sheep—"heritage breeds." That is because these animals have been selected to thrive on the scrubby vegetation and scarce water resources of the region over many centuries of care by Hispanic, Anglo, and Native American ranchers.

As the grinning farmer hinted to his visitors on that harvest day, heritage farming is in no way "retro," passive, or antiquarian; it dynamically adapts to new

Heritage Farm

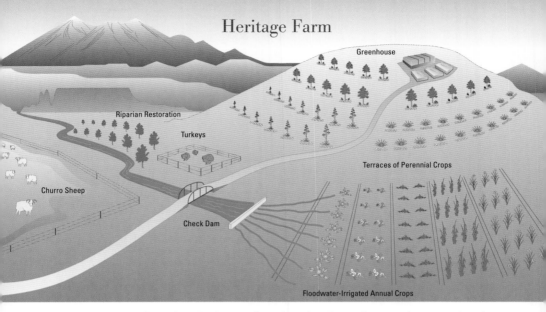

An artist's conception of a modern-day heritage farm that is based upon diversity of native seeds and animal breeds, and microhabitats. Note the check dam used to catch rainwater in a wash for irrigation use. This heritage farm has the prized churro sheep for wool and meat, and heritage turkeys for food and insect control in the vegetable garden.

opportunities through sustainable solutions that can stand the test of time. Rather than forgetting history, it incorporates the traditional knowledge orally passed from generation to generation within the region, just as the seeds of the crops themselves are passed hand to hand.

At the same time, the harvests from heritage farming are now reaching the market in ways that garner prices far higher than those which producers of commodity foods gain from their run-of-the-mill seeds and breeds. National and regional surveys show that consumers are now actively seeking

Heritage foods are increasingly featured at southwestern buffets.

out place-based foods and fiber products—from blue corn piki bread to Navajo weavings of Churro wool—that maintain the unique culinary and artistic legacies of the region. In recent years, growing concerns over food security, safety, and sustainability have swung public opinion in favor of locally produced heritage foods. One survey found that 90 percent of southwestern residents surveyed were willing to pay as much or more for heritage foods as they were for conventionally

Farmers markets have been established in many cities and towns across the Southwest as residents increasingly appreciate the taste and nutritious advantage of purchasing locally grown fruits and vegetables. In farmers markets such as this one in Flagstaff, Ariz., customers can find heritage varieties not generally available in supermarkets. Note the varieties of heritage tomatoes in the basket at the left.

produced commodity foods. Surprisingly, 57 percent said they were willing to pay 10 percent or more "than grocery store prices" to have access to such foods. Over half of the southwestern residents surveyed were willing to purchase place-based heritage foods as a means to support the rural traditions of the region's resident cultures.

And yet, newcomers to the region are perhaps the most eager to sample these foods. When foreign visitors come to the Grand Canyon or Mesa Verde, they wish to experience a taste of those special places, and locally produced heritage foods offer them that opportunity. Today, hundreds of thousands of people come to the Southwest to participate in the region's many saint's day feasts, American Traditions Picnics, chuckwagon dinners, chile cook-offs, "Flavors Without Borders" festivals, round-ups, and planting ceremonies.

Above: Navajo rugs have been bought and sold at Hubbell Trading Post for more than a century. Lower left: Vast varieties of chile peppers are traditional crops that remain a key part of southwestern cuisine. Lower right: Apple trees with a lineage dating back sometimes hundreds of years are being nurtured as part of the Kino Heritage Fruit Trees Project.

The Southwest—fondly known to some as Chile Pepper Nation—is widely recognized today as the region in North America where the most ancient agricultural traditions have survived and where the greatest crop diversity north of the tropics continues to be found in native gardens, fields, and orchards. It is not, however, the only region where a revival of support for heritage farming and place-based foods has gained momentum over the last decade. During that period, there has been a 22 percent rise in local food purchases nationally, and many of those vegetables, fruits, and meats are place-based foods derived from heritage farms and ranches. The number of farmers market outlets for such foods has tripled over the last 15 years to approximately 5,000, offering opportunities to heritage farmers and ranchers to sell their wares. For the first time since World War II, the annual sales of locally grown foods across the country is more than $5 billion.

At some places, like Hubbell Trading Post National Historic Site and Capitol Reef National Park, the National Park Service has played a catalytic role in supporting growth in the sustainable production, marketing, and educational outreach of place-based foods and handicrafts by featuring them in visitor centers and at festivals associated with the parks,

Heritage orchard in Slide Rock State Park, Sedona, Ariz.

monuments, national heritage areas, and historic sites on the National Register of Historic Places. That is because Congress specifically asked the National Park Service to explore viable ways to encourage the production of traditional products in national parks that "reflect, educate, and celebrate the unique history, spirit, culture, and natural treasures of the designated region and individual park."

From the Historic Fruita Orchard District in Capitol Reef to the Kino Heritage Fruit Trees Project at Tumacácori National Historical Park, you can "taste the traditions" of the Southwest. Although farming and ranching historically occurred on some lands now managed by the National Park Service, not all park visitors have had the chance to taste the fruits of sustainable, place-based food production that continues to be fostered in places as diverse as Canyon de Chelly National Monument, Great Sand Dunes National Park and Preserve, and

Dried gourds have been used in the Southwest for centuries as vessels for carrying water and storing seed, as well as for decoration.

Lyndon B. Johnson National Historical Park. The following narratives and photographs portray how these activities are part of the natural and cultural legacy of the Southwest. They also present ways to join in the celebration of these unique traditions at parks, monuments, historic farms, agricultural museums, farmers markets, and food festivals throughout the southwestern region. In short, heritage farming and ranching is good to think about, but its placed-based food products are just as good to eat.

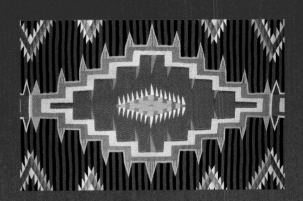

Left: Detail of historic Navajo Indian weaving

THE FLOWERING OF DIVERSITY
The Many Breeds
and Seeds of the Southwest

s the crow flies, it is not all that far from the Zuni waffle gardens around the Ojo Caliente wetlands in northwestern New Mexico to the Navajo orchards and dry-farmed corn fields perched high in the Chuska Mountains of northeastern Arizona. And yet, you might feel as though you were changing planets if you move too abruptly from the salt-adapted desert scrub surrounding Zuni gardens and springs, to the forests of Douglas fir and spruce high in the Chuskas. Within a two-hour drive, you can move through several of the "life zones" of natural vegetation first described by C. Hart Merriam a century ago, but in doing so, you would also be scaling a gradient of "crop hardiness zones" that define various constraints on crop production in the Southwest.

This landscape heterogeneity is the first clue to why there are so many food varieties found in the Southwest. As you move toward any horizon in this region of desert basin and rugged mountain range, you are likely to move up or down, not tens, but hundreds of feet in elevation. Your second clue to why the crops of this region are so diverse is its close proximity to Mexico, which is perhaps the oldest center for agricultural origins in the New World. Ancient seeds were easily carried up the trails

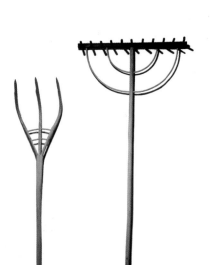

Above: Sunflowers, a heritage crop, are grown for the seed, which can be eaten or pressed for cooking oil.

Left: Native woods were fashioned into rakes, pitch forks, and other farm implements.

Oldest Surviving Heritage Farm in the Southwest

It is less than fifty miles north of the skyscrapers in the heart of metro Phoenix, but Table Mesa still tells a centuries-old story that cannot be drowned out by the street noise of the Phoenicians nearby. It is a story of one of the earliest domesticated crops in North America—a century plant now known as the Hohokam or Murphey's agave—that continues to grow on the very same cobble-lined terraces where it was first planted on the slopes of Table Mesa some 750 to 925 years ago. It may be the only case in North America confirmed by botanists and archeologists that the same genetic clone of a domesticated crop planted centuries ago has persisted in growing in the very same fields next to the very same harvesting tools that were used to manage it in prehistoric times.

Prehistoric pueblo-like structures have been found on many ridge tops associated with the nearby Agua Fria National Monument and on Table Mesa itself. Although the Table Mesa ruins have been badly looted, it appears that the fields below them were ignored by pothunters and amateur archeologists. Both the agave fields and the castle-like ruins in this central Arizona ecological transition zone are associated with a trade network between the upland ancestors of the Hopi and the lowland predecessors of the Salt and Gila River valleys.

Ancestral Puebloans built small, stone check dams to capture runoff water for crop irrigation.

On the ground, it is not hard to stumble across hundreds of yards of rock-lined terraces that follow the contours of Table Mesa. Here and there small clones of century plants poke up above the native vegetation, while oddly robust fishhook and cholla cacti also abound along the terrace lips. What is most telling is the presence of a stone tool only associated with the prehistoric harvesting of agaves in the Southwest—the so-called large agave knife with its edges unmistakably worked into a sharp blade by prehistoric flint-knappers. Here, within the flyway of Sky Harbor Airport and within a half mile of the constant hum of car and truck traffic on Interstate 17, lies America's oldest heritage farm.

Agave grow in the Santa Rita Mountains of southeastern Arizona. The Yavapi Indians of central Arizona utilized roasting pits to process cultivated agave for food and wove agave fiber.

from central Mexico into what is now the American Southwest. In other words, the Southwest has the greatest antiquity of crop cultivation of any region north of Mexico. From 2100 BC to AD 1540, some twenty cultures native to the region adopted a peculiar set of agricultural crops and practices; some originally diffused northward from Mexico, but others are undeniably unique to the varied landscapes of the Southwest.

Over the last four thousand years, indigenous and immigrant farmers have tried to grow various crop varieties in each of these zones, and have found microenvironments where some crops do better than others. By selecting particular varieties especially suited to the soils, rainfall, and growing season length in each of these zones, farmers of the Southwest now grow vegetables, grains, and fruits from below sea level in the Imperial Valley of southern California to well above 8,000 feet in the Chuska Mountains on the border of Arizona and New Mexico. This has prompted Southwest archaeologist Ann Woosley to suggest that prior to European conquest, indigenous farmers of the prehistoric Southwest farmed and gardened in a wider range of microenvironments and life zones than did all the other farmers in the rest of North America!

Ancestral Puebloans valued the fruit of the prickly pear cactus (nopales) and fashioned tongs from native trees to harvest the fruit.

Above: Prehistoric farmers grew several varieties of corn, and an increasing number of Southwest farmers today are showing renewed interest in heritage varieties. Inset: Tomatillos were brought up from Mexico centuries ago. Tomatillos, members of the nightshade family, are a staple in Mexican cuisine. Today, they are grown throughout the Southwest.

This mix of agricultural environments fostered the diversification of the "gene pools" of each of the food crops grown in the Southwest over the centuries. In trying to imagine this diversity, we must understand that Native Americans harvested more than crops of corn, beans, and squash. For starters, consider that

From prehistoric times, beans of many types have been a diet staple in the Southwest.

prehistoric farmers in the Southwest grew several very distinctive species of corn: chapalote, reventador, onaveño, 60 day flour/flints, and Puebloan long-cobbed flour/flints. Each of these came in several colors, with blue, red, and purple strains being adapted to colder hardiness zones. There was not just one bean species or variety, but many: tepary, jack, lima, and common pinto, each with a different maturation rate and tolerance to heat. There were several varieties for each of these bean species. In addition, three species of Mesoamerican squashes and pumpkins enlivened native gardens in prehistoric times, and a fourth arrived during the Colonial period. Some have developed tolerance to squash vine borers, while others repel flying insects that transmit viruses.

But corn, beans, and squash do not tell the entire story of the diverse legacy of heritage farming in the Southwest. Two species of grain amaranths, a domesticated lamb's quarter, sunflowers, a Sonoran millet, tobacco, tomatillos, gourds, cotton, agaves, and prickly pears originated in Mexico and were then taken north where they diversified after centuries of cultivation in this peripheral region. Insect control in these fields and gardens full of these crops was offered by domestic turkeys that were also traded up from Mexico.

Sooner or later, the farmers of the Southwest realized that they needn't wait for all their crops to come from what is now Mexico; they could freshly domesticate some themselves. They experimented with a little barley, several different agaves, wolfberry, devil's claw, beeweed, potatoes, bushmint, chiltepines, and several other so-called "field weeds." Some, like devil's claw, became fully domesticated, while others, like beeweed and cholla cacti, seem to have reverted to their wild status. Clonal descendants of certain agaves—while no longer cultivated for food today— can still be seen growing on the very same stone-lined terraces where they were

Devil's claw

intentionally planted and tended more than five centuries ago. This mix of fruits, roots, shoots, leaves, grains, and fowl offered prehistoric cooks a flavorful set of ingredients to prepare through boiling in clay pots, steaming in earthen pits, grilling on coals, or parching in baskets. Contrary to the popular but incorrect notion

that North America had few domesticated crops before Columbus and offered its inhabitants a meager menu from which to feed itself, it is clear that prehistoric farmers and chefs of the Southwest had dozens, if not hundreds, of food varieties from which to choose.

An illustration depicts the importance of corn to ancestral Puebloans in what is now the southwestern United States. Women are shown in a pit house making something similar to modern day piki bread. Inset: Navajo family members gather corn pollen for traditional ceremonial use.

Once the Spanish began to journey north out of central Mexico, they not only brought with them more annual crops, but also introduced livestock, honeybees, winter vegetables and grains, fruit and nut trees, as well as novel cooking techniques. Wheat, barley, lentils, peas, chickpeas, fava beans, melons, and watermelons were introduced into the region between 1590 and 1690, after which new varieties emerged that became unique to the Southwest. The Spanish also brought the fruits and rootstocks from Europe, Asia, Africa, and Mesoamerica to the region. For instance, apples became established around the Salinas Pueblo Missions as early as 1628, and these so-called *manzanos Mexicanos* still grow on the banks of streams flowing out of the Manzano Mountains east of Albuquerque today. In addition, Spanish priests and colonists offered their Native American neighbors peaches, figs, pomegranates, dates, olives, quinces, almonds, grapes,

pears, cherries, apricots, citrus, chiles, and other fruits. Churro sheep, Criollo-Corriente cattle, goats, chickens, and geese were intentionally brought in to the region, while European honeybees made it here on their own. New ways of baking, poaching, roasting, fermenting, and frying introduced new culinary possibilities. The Indo-Hispanic culinary traditions found today in the Southwest have deep and

Above left: What is thought to be the oldest quince tree in Arizona continues to grow on a ranch on the eastern slope of the Santa Rita Mountains in southeastern Arizona. Its age is estimated at well over 100 years. Below left and above: The Kino Heritage Fruit Trees Project in Arizona is successfully planting heritage apples and peach varieties. The apples, pictured lower left, are being grown in Patagonia and the clingstone peaches, pictured above, in Tucson.

diverse roots. Certain lamb and chickpea stews still made in northern New Mexico today can trace their origins back beyond Moorish Spain to Morocco, Arabia or Persia, and to central Asia as well. As New Mexican historian Juan Estevan Arrellano has explained, there are Persian, Arabic, Moorish, Greco-Roman, Mayan, and Aztec influences embedded in every southwestern garden and kitchen, so that terms like "Spanish" or "Indian" do not necessarily capture the ultimate origins of all the foods we find on the traditional southwestern table. While those extensive influences may be hard to cover in a casual conversation over the dinner table, they certainly make for wonderful eating!

Of course, the diversity of farming and foodways in the Southwest did not stop growing when the region became part of the United States. After Europeans, Africans, and Asians immigrated into these desert and mountain landscapes, native farmers borrowed and adapted additional foods and techniques from their new neighbors as well. These introduced foods included dates and olives brought from

northern Africa, as well as berries and cherries from northern Europe. They also included other breeds of livestock—from angora goats to Hereford cattle. Following the Gadsden Purchase in 1853, the region then known as the Great American Desert was broken into territories, and then into states of the United States of America. By the time Arizona and New Mexico finally attained statehood in 1912, the South-

Volunteers plant starts of quince trees.

west harbored the greatest diversity of heirloom crop seeds and heritage livestock breeds of any region in North America.

Today, the ancient threads woven into the network of heritage farming in this dry region remain more evident than they do anywhere else in North America, where native farming has largely been obliterated by industrial agriculture and urban development. We can witness the oddest of juxtapositions—some of the wildest landscapes left untrammeled on the entire continent, surrounding some of the most ancient and enduring agricultural landscapes. As they have for centuries, spring-fed gardens and orchards continue to sustainably produce food for residents of the Southwest in the midst of barren slickrock, shifting dry sands, and

75271
Johnson's Seedless Winter Pear
from J. B. Johnson
Box 98   Provo City Utah Co Utah
                                    D. G. Passmore
                                        2.25.98
                                    2-18-98

Provo City, Utah Co. Utah -

Quince

Top: A U.S. Department of Agriculture
drawing of quince in 1898
Left: An historic quince grove gone feral
in southeastern Arizona
Right: Mature San Rafael quince ripening
on the tree

indomitable wilderness. Traditional southwestern cuisine is now internationally celebrated, with foods resembling pinto bean *chimichangas* and blue corn *enchiladas* being offered in restaurants from Alice Springs, Australia and Almaty, Kazakhstan to Hamburg, Germany and Tokyo, Japan. But just where can you find foods unique to the Southwest being grown today as they have been for centuries, flowering and fruiting out of the arid southwestern soil?

Many restaurants in the Southwest now feature heritage foods on their menus. This meal, that included prickly pear cactus pads, tepary beans, heritage squash, and pumpkin seed sauce, was served to members of the Santa Cruz Valley Heritage Alliance in southeastern Arizona.

FARMING THE HERITAGE
Returning to the
Roots of Southwestern Agriculture

A miraculous revival of heritage farming is happening in the multicultural communities of the Southwest today. It involves the integration of drought-tolerant heirloom seeds and trees, heritage breeds of livestock, and traditional farming techniques with modern technological advances and the creative marketing of place-based foods. There are heritage farms cropping up from Patagonia, Arizona to Taos, New Mexico, and many of the farmers managing them are making more profits now than they have for years. But to understand why this revival is truly a miracle, it is important to realize how dramatic the decline in food production has been in the Southwest over the last half century.

The great pueblos, such as those in Chaco Canyon, New Mexico, were eventually abandoned when, many think, the water supply became insufficient to sustain vital agriculture.

Members of the Tohono O'odham Nation harvest tepary beans on their reservation south of Tucson, Ariz., using both hand tools and sophisticated machinery. Inset upper left: Hopi women sort and clean wool from locally grown Churro sheep. Inset lower left: A bin of dried gourds, a traditional crop both useful and decorative. Inset below: Avelisto Moreno, left, and Cedric Norris harvest watermelons at San Xavier Cooperative Farm in Arizona.

What looks like the arrival of vital rain at sundown in Tucson, Ariz., is just a beautiful, but insignificant, virga, rainfall that evaporates before reaching the ground.

Farmers in the Southwest have always had to be careful about the amount of water, electrical power from fossil fuels, and gasoline they use in producing food, but the overuse of these resources is now a critical issue facing all of society in this dry, dusty landscape. As a result of global warming, long-term drought, and declining fossil fuel supplies, there just isn't enough water and fossil fuel to go around if everyone tries to do their business as usual.

A half century ago, farmers and ranchers in the region used 90 percent of all water diverted out of rivers, lakes, and reservoirs in the Southwest, simply because there were few competing uses. They had all but abandoned many of the drought-

hardy, heat-tolerant heritage crops of their forefathers because they believed that fossil fuel and water would never again be scarce commodities. Ironically, the post-World War II migration to the Sun Belt—one of the most massive migrations in all of human history—suddenly pitted their agricultural use of water against the residential, recreational, and industrial needs of other users.

Farming itself—as a livelihood that can be passed down from one generation to the next—has become endangered. Every minute, two acres of American farmland are taken out of cultivation for development, and in the Southwest, that rate is occurring 50 percent faster than it did a quarter century ago. Until recently, heritage farming fared no better in its survival than did other forms of farming. For example, between 1950 and 2000, the number of farms and ranches in Arizona dropped from 11,400 to 7,300, a decrease of 36 percent, resulting in a 37 percent decrease in acreage from 45.50 million to 26.57 million acres. Farmers and ranchers in neighboring states were in the same sinking boat. Nevada and New Mexico are among the states with the most rapidly accelerating rates of farm and ranch land loss. The American Farmland Trust lists ten counties in the Southwest on their "red list" of places where prime farm and ranch land sits most

precariously in the path of development: Five counties in Utah as well as Colfax and Rio Arriba counties in New Mexico, Saguache and Montrose counties in Colorado, and Pinal County in Arizona.

As the fossil fuel costs of pumping water from aquifers or reservoirs skyrocketed after the energy crisis of the 1970s, many farmers in the Sun Belt found ways to improve their water-use efficiency per acre. They cut their overall water use per harvest by a third over the last quarter century, becoming ever more thrifty. Farmers and ranchers now realize that they must be innovative in order to find ways of reducing their inputs merely to survive; if they were to stand still, they would surely get left behind.

It is now evident to society at large that farming families who have stewarded the same land for several generations are simply not ready to throw in the towel and let their land be engulfed by tumbleweeds or subdivisions. Many of them remember how resilient their ancestors were during other times of stress—the Dust Bowl era immortalized by John Steinbeck in *The Grapes of Wrath*, or the fifties drought described so well by Larry McMurtry in *The Last Picture Show*—and have found ways to use their own families' traditions to both restore productivity to their lands and to tell the story of why heritage farming is so important today.

Some have stayed on the land by returning to the very same drought-hardy heirloom seeds and heritage breeds that served their predecessors so well during

Left: Tobacco was brought to the Southwest from Mexico.

Above: A stand of corn in Ganado, Ariz., where an historic dam and canal system are being restored to irrigate area fields

Right: An apricot tree laden with fruit has been selected for its origins in the Spanish-mission era by members of the Kino Heritage Fruit Trees Project.

Inset: Churro wool is weighed on a ranch in northern Arizona.

Above: Historically, hay grown in the fields around Hubbell Trading Post has helped support livestock raised at the post and throughout northeastern Arizona.
Below: Heritage beans, harvested in northern Arizona, are dried in a basket woven from native plants.

other eras. Thus, they are not merely growing and marketing food, but are growing stories about those foods that can help to market their products and educate consumers as well. They are not merely producing calories, but they are also making history.

HUBBELL TRADING POST
National Historic Site: Ganado, Arizona

H

ubbell Trading Post, hidden along the riparian growth of a wash just east of the Painted Desert, has been a safe haven and crossroads for well over a century. Within a decade of the Navajo people returning from the "Long Walk" to Bosque Redondo in New Mexico, trader John Lorenzo Hubbell found enough families in the Ganado, Arizona area to encourage him to establish a trading post there. It has since become the longest continuously operated trading post on Navajo Nation lands, and the most widely visited by admirers of the world-renowned Navajo rugs. The rugs are best appreciated in the context that Hubbell began his trading in northern Arizona just as the Navajo-Churro sheep were recovering in number from government-sponsored reductions, and the Navajo themselves were entering into the larger cash economy. Not only were orchards, gardens, pastures, and fields integrated into Hubbell's own homestead at this historic site, but Navajo families from miles around brought

Above: Pastures, gardens, fruit trees, and livestock still play
an important role at Hubbell Trading Post today.

John Lorenzo Hubbell

their produce, wool, and piñon nut harvests to Hubbell for sale or barter.

Beginning in 2005, the National Park Service staff at Hubbell Trading Post have been engaging both the surrounding Navajo communities and heritage farming experts in plans to revive the farming economy of the Ganado Valley. A newly constructed reservoir in the watershed above the trading post offers opportunities for Navajo families to undertake irrigated farming of foods and livestock forages once again. The former fields, pastures, and orchard areas at Hubbell are also being replanted once more, using water from the reservoir to demonstrate sustainable farming techniques. These are being replanted with the very same local varieties and heirlooms which John Lorenzo Hubbell himself ordered out of seed and nursery catalogs from a century ago. Navajo-Churro sheep graze seasonally at the site. More importantly, perhaps, National Park Service staff and volunteers support the nonprofit conservation and education organization known as Diné Be' iiná, which translates as "Sheep is Life" or "Navajo Lifeways."

Left and above: In the late 1800s, goods arrived and departed by covered wagon or on horseback. The trading post continues today to be the premier outlet for ornate traditional and modern woolen rugs woven by Navajo weavers. Inset rug photos from the top counterclockwise: Teec Nos Pos-style rug by Bessie Lorgo of Arizona; Yei-themed rug by Minnie Coan, Red Rock, N. Mex., and Two Grey Hills rug by Maralin John of Two Grey Hills, N. Mex.

NAVAJO FIELDS AND ORCHARDS
Associated with Canyon de Chelly
National Monument: Chinle, Arizona

The peaches of Canyon de Chelly echo the colors of the cliffs above. Navajo families, whose ancestors may have farmed and herded in the canyons of the Chuska Mountains for the last three centuries, are the prime managers of fields and orchards in Canyon de Chelly and adjacent Canyon del Muerto. These traditions are interpreted by Navajo tribal members on the National Park Service staff based at a visitor center near Chinle, Arizona. The real work of heritage farming, however, is accomplished by more than twenty families who actively raise field crops, fruit trees, and livestock back in the far reaches of Canyon

Canyon de Chelly, above and right, and adjacent Canyon del Muerto in northeastern Arizona have supported Navajo sheep herders and small-scale agriculture for hundreds of years. The petroglyph of a corn stalk (inset) is evidence of the importance of a corn plant in the present-day Southwest before the arrival of European explorers in the 1500s. This petroglyph was photographed during the Chicago Field Museum McCormac Hopi Expedition in 1901.

del Muerto and Tsegi Canyon. The Navajo families in the canyon may be the third or even fourth culture to farm the floodplains of these canyons since corn reached the Chuska range some 2,500 years ago, but some of their varieties and techniques are much the same as those used prehistorically. Others, such as the planting of fresh pits of Spanish-introduced peaches, apricots, and other fruits, are said to have begun three to four centuries ago. Small but delicious cream-colored clingstones and freestone peaches were commonly planted and harvested in these canyons since the first orchards were established near White House Ruin.

While historical accounts reveal that roughly four thousand of these so-called "Navajo peach trees" were cut down by U.S. army forces led by Capt. John Thompson and Kit Carson in 1864, supposedly leaving the Navajo with no remaining trees, oral history among the Navajo themselves suggest that examples

> Roughly 4,000 "Navajo peach trees" were cut down by
> U.S. Army forces in 1864,
> but some of the original stock survived.

of this original stock survived in the deepest recesses of the Chuskas where soldiers failed to tread. However, the droughts of the last century no doubt took their toll on most of those survivors. In the interim years, Navajo families received additional cultivars (varieties under persistent cultivation) of peaches and other fruits from many sources, but Mormon introductions from Utah were especially commonplace. At the same time, Navajo farmers have successfully safeguarded their historic heirlooms of corn, beans, squash, and other annual crops without any assistance needed from the National Park Service or the Navajo Nation Department of Agriculture. Canyon de Chelly residents have participated in heirloom fruit tree and historic orchard restoration workshops associated with the Southwest Regis-Tree project, and aspire to initiate planting of historic fruit tree stock from other parts of the Navajo Nation.

Navajo farms in Canyon Del Muerto, left, in northeastern
Arizona as seen from the air in an undated photo prior to 1979

CAPITOL REEF NATIONAL PARK
Fruita Rural Historic District:
Torrey, Utah

The first orchards and fields planted below the red rock cliffs in what is now Capitol Reef National Park were established in 1882. In that year, Nels Johnson, a Scandinavian convert to the Mormon faith, made claim to 160 acres in what soon became the village of Fruita, Utah. By the late 1880s, the "Dixie" Mormon farmers of Fruita became infatuated with the quality of fruit tree varieties offered by the Stark Brothers Nursery in Louisiana, Missouri, and invited one of the brothers, Clarence Stark, to visit them. Thus began a long-term relationship between Fruita orchards and the Stark Brothers Nursery, which provided them by rail with dozens of fruit tree varieties never before evaluated in the Southwest. Over the next seven decades, the farmers of Fruita planted at least twenty-five Stark Brothers selections of apples, apricots, cherries, grapes, peaches, pears, pecans, plums, quinces, and black and Carpathian walnuts, many of them now considered to be historically important heirlooms. One of the most interesting remnants from these original Mormon plantings is a native plum known as the Potawatomi, which occurs west of the Rockies only as hedges and feral shrubs in Mormon farming villages. Mormon elders in rural towns not far from Capitol Reef recall how this plum literally served as their survival food during the Dust Bowl of the Great Depression, when families were too poor to import food from other regions.

Today, twelve of the seventeen historic orchards scattered along the floodplains of the Fremont River and Sulfur Creek are being restored to their historic context

Above left: Drying peaches in Canyon Del Muerto in northeastern Arizona sometime in the mid-20th century. Below left: A heritage orchard in Fruita, Utah. Inset: A ripening pomegranate in Arizona.

The heritage orchard and majestic butte in historic Fruita, Utah, Capitol Reef National Park

Many varieties of fruit and nut trees are now

considered to be historically important heirlooms.

Above: Elizabeth Berry's Gallina Canyon Ranch in Abiquiu, N.Mex., has specialized in growing heirloom crops since 1986. The amaranth, corn, chilies, squash, tomatoes, and beans are sold to restaurants in the Santa Fe area that boast menus featuring heritage foods. Berry grows 30 varieties of tomatoes and 98 kinds of beans, including White Aztecs. The heritage produce is popular at a weekly farmers market in Santa Fe. Inset: Visitors to Capitol Reef National Park in Utah are invited to harvest fruit from the heritage orchards in the historic Mormon settlement of Fruita. The National Park Service maintains approximately 2,600 fruit trees within the park's boundaries.

prior to World War II, with the inclusion of modern integrated pest management techniques to reduce damage to trees and their fruit yields. A total of 2,654 fruit trees and vines, representing sixty-five varieties of twelve fruit and nut species, have been identified, tagged, and mapped. A "you-pick" harvesting opportunity in late summer and fall still attracts hundreds of harvesters. The heirloom fruits of Fruita are prominently featured on the menu of a nationally renowned "local foods" restaurant in nearby Boulder, Utah.

TUMACÁCORI NATIONAL HISTORICAL PARK
Tumacácori, Arizona

*Heritage farming is making a comeback at
Tumacácori National Historical Park.*

S ome thirty miles west of Patagonia lies one of the oldest Spanish missions
still standing in Arizona; its mission church walls rise above the riparian
vegetation, fields, and orchards along the Santa Cruz River. For centuries,
the Tumacácori Mission was more than a sturdy and beautiful Catholic church;
it was the focal point for rural life in the upper Santa Cruz River watershed, and
once included a five-acre orchard of Spanish-introduced fruit trees. By the 1990s,
only one stately tree of the mission fig still stood in the historic garden behind
the visitor center, and National Park Service employees eagerly worked with
horticulturists to preserve and propagate it. Then, in 2004, Tumacácori National

Historical Park staff invited botanists from the Arizona-Sonora Desert Museum and the University of Arizona to establish the base for the Kino Heritage Fruit Trees Project on newly acquired park lands. This project, named for Padre Eusebio Francisco Kino, who established the mission at Tumacácori in 1691, is attempting to locate, identify, and re-propagate all of the Spanish-introduced fruit trees of the Kino era (1687–1711). Kino and his Jesuit colleague, Juan de Ugarte, personally introduced many of the European fruits, nuts, seeds, and livestock breeds that later became the foundation for heritage farming in the Southwest.

In late spring of 2007, their dream became reality: roughly an acre of Sonoran quinces, pomegranates, mission figs, and other Kino-legacy heirlooms were planted on the grounds of Tumacácori National Historical Park as part of the National Park Service's efforts to preserve the living history of the region's farming traditions. Dozens of people helped plant the trees within sight of the historic mission church one morning, and hundreds more joined in the festivities. Within a matter of years, it is estimated some fifty trees will bear fruit and offer Tumacácori visitors a sweet taste of history.

The legacy of Father Kino is manifested in the Kino Heritage Fruit Trees Project, which includes the reintroduction of pomegranates and mission figs. Tumacácori National Historical Park Resources Manager Jeremy Moss is shown with a mission fig tree start during the dedication of the heritage orchard at the park. Lower right: A bronze sculpture of Father Kino stands in Statuary Hall in the Capitol in Washington, D.C.

Eusebio Francisco Kino

Native Seeds/SEARCH Conservation Farm
Patagonia, Arizona

I n 1997, Native Seeds/SEARCH, a nonprofit organization, began to develop its conservation farm for heirloom crops on the floodplain of Sonoita Creek, which runs through the colorful Hispanic-American village of Patagonia in southern Arizona. It is adjacent to, and in concert with, a Nature Conservancy conservation easement running along the Sonoita Creek Corridor. The Native Seeds/SEARCH Conservation Farm encompasses some sixty acres, where as many as 250 different heirloom seedstocks may be planted in any season. Native Seeds/SEARCH conserves, increases, and distributes heirloom vegetables, grains, and beans from Hispanic, Anglo, and Native American cultures of the bi-national Southwest, and the organization now plays a key role in the revival of

Native Seeds/SEARCH's 60-acre farm in Patagonia, Ariz.

heritage farming in the region. Prior to the organization's founding in 1982, few of these seedstocks were available to farmers and gardeners. While maintaining a seed bank of more than 1,200 accessions of annual crops, the staff also raises heritage breeds of turkeys and propagates heirloom fruits and nuts. They experiment with intercropping native grains and beans, and with using them in low-water-requiring permaculture designs. Permaculture is an agricultural system or method that seeks to integrate human activity with natural surroundings to create highly efficient, self-sustaining ecosystems.

Groups can visit the Conservation Farm during seasonal festivities such as the San Juan's Day celebration. This celebration has revived a historic southern Arizona tradition of cleaning ditches and offering prayers at dawn on the Feast of Saint John the Baptist, the patron saint of irrigation agriculturists. San Juan's Day is also the legendary beginning of the summer rainy season in southern Arizona and northern Sonora, Mexico.

Left top: Yellow watermelons, grown in Arizona, are prized for their honey-like sweetness. Left center: The Spring Festival at El Rancho de las Golondrinas in Santa Fe, N.Mex., includes a priest with a bulto of San Isidro blessing of the water, the fields, and the animals. A bulto is a small statue, and San Isidro is the patron saint of the ranch and agriculture. Left bottom: Blossoms indicate a bean crop is in the offing. Below left: Amy Schwemm sells her hand-made mole powders at the Santa Cruz Valley Foodways Festival in Arizona. Moles are spicy sauces of Mexican origin, usually having a base of chilies, onions, nuts or seeds, and unsweetened chocolate and served with meat or poultry. Below, right: Verna Miguel uses a colander and an electric fan to remove chaff from newly harvested tepary beans.

El Rancho de las Golondrinas
Santa Fe, New Mexico

As you come onto the high plains just south of Santa Fe, New Mexico, you may be able to spot El Rancho de las Golondrinas, a historic farm and living history museum nestled into the La Cienega Valley. The original colonial buildings on the farm date from the early eighteenth century and other historic structures from rural New Mexico built in the eighteenth and nineteenth centuries have been moved to the farm. These Colonial Era buildings are scattered across some two hundred acres of historic farmland that is once again green with traditional crops.

The museum opened its gates in 1972 to interpret the rural heritage and agrarian culture of Spanish Colonial New Mexico. In addition to growing both native and Hispanic field crops, as well as culinary and medicinal herbs, staff use draft horses to plow the land and Hispanic blacksmithing skills are featured. Special festivals and theme weekends offer visitors an in-depth look into the foodways, farming traditions, celebrations, and artisanal skills associated with rural life when this part of the Southwest was ruled by Spain and Mexico. The museum has twice hosted the annual national meetings of the Association of Living Historic Farms and Agricultural Museums, featuring the food and farming folkways of the Santa Fe Trail.

Above left: A young visitor at El Rancho de las Golondrinas in Santa Fe, N.Mex., with the help of an adult learns the art of plowing behind a horse. Corn, chiles, and other crops grow in the background at the heritage farm. Below left: A donkey powers the press used to crush sorghum to be rendered into molasses during the El Rancho de las Golondrinas Harvest Festival in Santa Fe, N.Mex. Right: The smithy made iron horse shoes and fashioned or repaired equipment in a shop like this at Fort Union, N.Mex.

An artist's conception of the pueblo of Pecos during the Spanish missionary period illustrates that farming and animal husbandry practices were well established at the time.

Colonial-era buildings are scattered across 200 acres of historic farmland

at the present-day Pecos National Historical Park in New Mexico.

Slide Rock State Park
Pendley Historic Orchards
Sedona, Arizona

In one of the most stunning settings in Sedona, Arizona, where thousand-foot crimson cliffs tower above, stands the remnants of an heirloom orchard. Here, on a narrow corridor of floodplain lands below the red rocks of Oak Creek Canyon, an Arizonan named Frank Pendley filed claim in 1910. Within two years he had planted his first apple orchard there. Today, one of those original plantings—an Arkansas black heirloom apple—still survives in Slide Rock State Park, where it is known as the "Heritage Tree," and is honored by inclusion on the National Historic Register. The Pendley family dry-farmed squashes and melons alongside their small apple orchard before completing an irrigation canal that brought water to their homestead from several hundred yards up Oak Creek.

The Pendley orchards then expanded to include at least fourteen other kinds of apples, including the rare Gano heirloom taken from a native tree planting in

The heritage orchard and apple-polishing machine at Slide Rock State Park

Missouri in 1834. The Pendley orchards also harbored apricots, blackberries, black walnuts, English walnuts, peaches, plums, pears, and strawberries, many of which still remain within the state park. In addition, annual crops such as sweet corn, watermelons, squash, carrots, tomatoes, and cabbage have been grown either by the Pendleys or by various Hopi farmers visiting the site over the decades. Today, the surviving heirloom fruit trees are being re-propagated on site, and the historic trees are being pruned, stabilized, and mapped. Inside the recently restored apple-packing shed, an old-fashioned apple polisher has been renovated, and is featured at Slide Rock's annual apple festival that is usually held one of the first two weekends in October. The Pendley Orchard is a classic example of the prime microhabitats that Anglo settlers selected for their homestead orchards from the 1880s onward in the canyon country of the Southwest.

HIGH DESERT FARMERS
Gallup, New Mexico

In a stretch of the Rio Puerco valley, slung between the Zuni and the Navajo reservations, artist-turned-farmer Steve Heil has pioneered perennial permaculture techniques to sustainably produce a number of culinary and medicinal crops. He grows them on drip irrigation below slickrock cliffs on sandy soils in a place that the Navajo call Tse Yaaniichii, "Where the Red Rock Ends." The ancestral Puebloans probably grew their native crops on these same soils some thousand years ago. Today, Steve Heil uses the help of volunteers and students to grow a number of healthful perennials such as Indian ricegrass, which makes a nutritious flour, and greenthread (*cota* in Spanish), which is one of the most popular Indian teas in the Southwest.

Because of his success in finding novel means of propagating greenthread as a perennial crop, and sharing his sustainable production techniques with others who now produce it commercially, greenthread was recently added to the Slow Food Ark of Taste as a rare food crop that is on its way to market recovery. Farmer Steve Heil, also a teacher in the McKinley County, New Mexico schools, offers visitors to the High Desert Farm an explanation of how perennial polycultures (mixed cropping) has the potential to slow the soil erosion and nutrient loss that has occurred in the Puerco watershed during decades of inappropriate grazing and monocultural crop production. Heil is demonstrating that farming of heritage foods is not only sustainable, but profitable enough to attract others to this endeavor.

Left: A docent at El Rancho de las Golondrinas Living History Museum in Santa Fe, N.Mex., strings chilies into ristas, arrangements of peppers to be dried for decoration or future consumption. Right: Steve Heil in his field of greenthread in New Mexico.

Above: Modern-day Navajos prize the high quality and naturally varied color of the wool from their flocks of Churro sheep. Above left: Herding Navajo-Churro sheep near Hard Rock, Ariz., with the help of three, well-trained sheep dogs. Below: A Navajo artisan begins weaving a rug on her loom.

The Remarkable Survival of Navajo-Churro Sheep

The oldest breed of domestic sheep still found in North America, the Navajo-Churro sheep nearly blinked out of existence a quarter century ago, with fewer than five hundred individuals left on the entire continent. Today, however, more than five thousand Navajo-Churro sheep are registered, and many of the ewes among them are bearing lambs year after year. For the first time in decades, Navajo-Churro lamb and mutton appears regularly on menus of fine restaurants in both Arizona and New Mexico. Their wool commands high prices among specialty weavers, and Navajo rugs made of 100 percent Churro wool now grace the walls of several galleries. This revival requires the concerted effort of many people to ensure that this breed gets the respects that it needs and deserves.

Some of the first livestock to be brought to the Americas, a herd of three thousand Churro sheep was driven northward out of central Mexico into the Rio Grande watershed by Onate's expedition in 1598. There, they were shared with Pueblo Indian farmers who tried out Spanish-style weaving, but abandoned these animals around the time of the Pueblo Revolt of the 1680s.

After a few more decades, they were permanently adopted by both Puebloan and Hispanic farmers of northern New Mexico, where Navajo families acquired them through barter or theft as early as 1704. By 1750, the Navajo tended enough sheep in the Chuska Mountains that overgrazing destroyed most of the grass and shrubs below the conifers, and wildfires became far less frequent in this country. Between 1820 and 1846, three million head of Navajo-Churros roamed the Colorado Plateau and their many uses deeply influenced the lifeways of the people there. From that point on, Navajos were prone to explain to outsiders that for their families, "Sheep is Life." The Churros of Spain had adapted to the dry conditions of scrubby forage of the Colorado Plateau and thus became known as a distinctive breed.

Nevertheless, during the years leading up to the Civil War, and again during the Dust Bowl, the federal government mandated stock reductions in Navajo lands that nearly put an end to this breed. It was not until the early 1980s that a number of farsighted livestock scientists, weavers, hobbyists, and community development activists revived the breed both among Hispanic and Navajo communities. Navajo-Churro sheep are now internationally recognized through the Slow Food Foundation for Biodiversity, and are on the way to full market recovery.

AFTERWORD

An emerging strategy for supporting the production and marketing of more place-based heritage foods is through a designation known as National Heritage Areas, which are aimed at aiding tribal and private land owners in their efforts to accomplish historic preservation, cultural revivals, and ecological restoration in great American landscapes. There are now National Heritage Areas designated in the Northern Rio Grande watershed of New Mexico, in the Mormon Pioneer communities of southwestern Utah, and at the Yuma Crossing of the Colorado River in Arizona. Others are proposed for the Santa Cruz River Valley of southcentral Arizona and the Little Colorado River Valley of Arizona and New Mexico. Through funds passed onto them through the National Park Service, farmers and ranchers in these areas are finding novel ways to promote and support the production of place-based foods derived from heritage farms and ranches. Despite a few storm clouds here and there, the future looks bright for heritage farming.

Above: Produce grown locally in the Santa Cruz River valley, including heritage foods such as prickly pear pads and fruit of the cactus, are featured in a meal in a restaurant operated by the Tohono O'odham Nation in southeastern Arizona.

FURTHER READING

Arrellano, Juan Estevan, ed. 2006. *Ancient Agriculture: Roots and Application of Sustainable Farming* by Gabriel Alonso de Herrera. Ancient City Press, Santa Fe, N.Mex.

Dahl, Kevin. *Native Harvest: Authentic Southwest Gardening*. Western National Parks Association, Tucson, Ariz. 2006.

Davidson, George E. *Red Rock Eden*. Capitol Reef Natural History Association, Torrey, Utah. 1986.

Diamant, Rolf, Jefferey Roberts, Jacquelyn Tuxill, Nora Mitchell, and Daniel Laven. *Stewardship Begins With People: An Atlas of Places, People, and Handmade Products*. Conservation and Stewardship Publication No. 14, Conservation Study Institute of the National Park Service, Woodstock, Vt. 2007.

Dunmire, William. *Gardens of New Spain: How Mediterranean Plants and Foods Changed*. America University of Texas Press, Austin, Tex. 2004.

Frank, Lois Ellen. *Foods of the Southwest Indian Nations*. Ten Speed Press, Berkeley, Calif. 2002.

Imhoff, Dan and Roberto Carra. *Farming the Wild: Enhancing Biodiversity of Farms and Ranches*. Sierra Club Books, San Francisco, Calif. 2003.

McLeod, Judyth. *Heritage Gardening*. Simon and Shuster, N.Y. 1994.

Nabhan, Gary Paul, ed. 2008. *Renewing America's Food Traditions*. Chelsea Green Publishing, White River Junction, Vt.

The author encourages consumers to purchase foods produced from historic seeds and breeds directly from Native American communities when available.

Copyright © 2010 by Gary P. Nabhan
ISBN: 978-1-58369-121-2
Published by Western National Parks Association

The net proceeds from Western National Parks Association
publications support educational and research programs in the
national parks. Receive a free Western National Parks Association catalog,
featuring hundreds of publications.
Email: info@wnpa.org or visit www.wnpa.org.
Written by Gary Nabhan
Edited by Dan Stebbins and Melissa Urreiztieta
Designed by Boelts Design
Printed by: C&C Offset Printing Co., Ltd.
Printed in: China
Photography: Cover: Josh Schachter, (inset) Lois Ellen Frank; title page: John Schachter; page 2: Lois Ellen Frank; page
3: courtesy of Albuquerque Biological Park, Rio Grande Botanic Garden; page 5: Chris Hinkle/Explorer Newspapers,
(inset) Timothy Willms/Talus Wind Ranch, Galisteo, N.M.; page 6: (top) Tom Bean, (inset) Chris Hinkle/ Explorer
Newspapers, (below) Scott Aldridge/Western National Parks Association; page 7: Amy Haskell, (inset) Geo. Huey; page
8: Tom Bean; page 9: (top and lower right) Tom Bean, (lower left) Chris Hinkle/Explorer Newspapers; page 10: Randy
Metcalf; page 11: Tom Bean; page 12: (top) photo courtesy of the National Park Service, (lower left) Scott Aldridge/
Western National Parks Association, (lower right) Jesus Garcia/Kino Heritage Fruit Trees Project; page 13: photo ©
Arizona State Parks; page 14: (top) Amy Haskell, (bottom) George H.H. Huey; page 15: (top) Scott Aldridge/Western
National Parks Association, (below) Efraín Padró; page 16: George H. H. Huey; page 17: (top) Bruce Griffin, (bottom,
left) George H. H. Huey; page 18: (top) Tom Bean, (inset) Walter Jeffries/Sugar Mountain Farm, Vermont, (below)
Native Seeds/SEARCH; page 19: Amy Haskell; page 21: (inset) Teresa Showa; pages 22-25: Jesus Garcia/Kino Heritage
Fruit Trees Project; page 26: Randy Metcalf/photographed courtesy of the Agave Restaurant, Desert Diamond Casino,
Tucson, Ariz.; page 27: photo courtesy of the National Park Service; pages 28-29: Chris Hinkle/Explorer
Newspapers, (insert upper left) Tom Bean, (inset lower left) Amy Haskell, (inset lower right) Josh Schachter; pages 30-
31: Warren Faidley; page 32: (above) photo courtesy of American University, (below) Scott Aldridge/Western National
Parks Association; page 33: Jesus Garcia/Kino Heritage Fruit Trees Project, (inset) Tom Bean; page 34: (above) historic
photo courtesy of the National Park Service, (below) Tom Bean; page 35: George H.H. Huey; pages 36-37: historic
photos courtesy of the National Park Service, (rugs) George H.H. Huey; pages 38-39: (top) Joel Grimes; page 39:
(lower right) historic photo courtesy of the National Park Service, (inset) Sumner W. Matson, Chicago Field Museum's
McCormick Hopi Expedition, courtesy of the Milwaukee Public Library; page 40: ©Laura Gilpin, 1981, Laura Gilpin
Collection, Amon Carter Museum, Fort Worth, Texas; page 42: (top) ©Laura Gilpin, 1981, Laura Gilpin Collection,
Amon Carter Museum, Fort Worth, Texas, (below) Efraín Padró, (inset) Jesus Garcia/Kino Heritage Fruit Trees Project;
pages 44-45: Efraín Padró; page 46: Lois Ellen Frank, (inset) from an Internet posting, photographer unidentified; page
47: Josh Schachter; page 48: Jesus Garcia/Kino Heritage Fruit Trees Project, (inset left) courtesy of Jeremy Moss, (inset,
right) U.S. Department of Agriculture, (inset below) photographer unknown, sculpture by Suzanne Silvercruys/Statuary
Hall, U.S. Capitol; page 49: courtesy of Native Seeds/SEARCH; page 50: (top and bottom) Amy Haskell, (center)
Shirley Barnes; page 51: (left) Vanessa Bechtol/Santa Cruz Valley Heritage Alliance, (right) Josh Schachter; page 52:
(top and bottom) Shirley Barnes; page 53: George H.H. Huey; pages 56-57: photos © Arizona State Parks; page 58:
Josh Schachter, (inset) Shirley Barnes; page 59: photo courtesy of Steve Heil; page 60: sheep photos by Tom Bean, (below)
Western National Parks Association; page 62: Randy Metcalf/photographed courtesy of the Agave Restaurant in the
Desert Diamond Casino, Tucson, Ariz.; back cover: Tom Bean.

Illustrations: Page 3: Dan Boone, Ron Redsteer, Gary Nabhan, courtesy of Northern Arizona University; pages 20-21,
Roy Andersen, courtesy of the National Park Service; page 25: (top) U.S. Department of Agriculture; pages 54-55: Roy
Anderson, courtesy of the National Park Service

Cover: A pumpkin grown in Arizona serves as the background for an assortment of vegetables, fruit, and herbs from
New Mexico farms.
Title page: Anthony Moreno harvests corn on the San Xavier Cooperative Farm in Arizona's Santa Cruz Valley.